Off-Grid Guide:
Live Off The Grid With DIY Micro Hydro Power System

Disclaimer: All photos used in this book, including the cover photo were made available under a <u>Attribution-ShareAlike 2.0 Generic (CC BY-SA 2.0)</u>

and sourced from <u>Flickr</u>

Copyright 2016 by Publisher – All rights reserved.

This document is geared towards providing exact and reliable information in regards to the topic and issue covered. The publication is sold with the idea that the publisher is not required to render accounting, officially permitted, or otherwise, qualified services. If advice is necessary, legal or professional, a practiced individual in the profession should be ordered.

- From a Declaration of Principles which was accepted and approved equally by a Committee of the American Bar Association and a Committee of Publishers and Associations.

In no way is it legal to reproduce, duplicate, or transmit any part of this document in either electronic means or in printed format. Recording of this publication is strictly prohibited and any storage of this document is not allowed unless with written permission from the publisher. All rights reserved.

The information provided herein is stated to be truthful and consistent, in that any liability, in terms of inattention or otherwise, by any usage or abuse of any policies, processes, or directions contained within is the solitary and utter responsibility of the recipient reader. Under no circumstances will any legal responsibility or blame be held against the publisher for any reparation, damages, or monetary loss due to the information herein, either directly or indirectly.

Respective authors own all copyrights not held by the publisher.

The information herein is offered for informational purposes solely, and is universal as so. The presentation of the information is without contract or any type of guarantee assurance.

The trademarks that are used are without any consent, and the publication of the trademark is without permission or backing by the trademark owner. All trademarks and brands within this book are for clarifying purposes only and are the owned by the owners themselves, not affiliated with this document.

Table of Contents

Live Off The Grid With DIY Micro Hydro Power System 1
Introduction 4
Chapter 1 – Getting Started 5
Chapter 2 – Micro Hydro System Sizing & Components 9
Chapter 3 – Micro Hydro Power Generator 14
Chapter 4 – Paddle Wheel Hydro Generator 18
Chapter 5 – Recycled Hydro Power Generator 22
Conclusion 26
FREE Bonus Reminder 27

Introduction

Going off-grid and living a self-sustaining lifestyle isn't as tough as it sounds. These days, lots of people are leaving their hectic lifestyle and going off-grid. If you have the same thought in mind, then you have certainly come to the right place. With a little planning, you too can live on your own in a prosperous way.

We all are dependent on government supplies for almost every major resource. From electricity to food, the supply to almost every basic necessity of ours is controlled by someone else. In the present scenario, where the future is uncertain, we need to take a full control of our lives.

You can easily do this by going off-grid. Not only will you be able to survive an unforeseen scenario, but you can also save big on your pocket. Additionally, you can go green and utilize various natural resources while diminishing your carbon footprint. In order to go off-grid, you need to have a constant and reliable supply of power. Without it, you won't be able to keep your house warm/cool or prepare your food.

There are plenty of ways one can harness natural resources to generate power. In this comprehensive guide, we will teach you how to utilize hydro power and create your own micro hydro power generator. We will provide a stepwise solution for you in this DIY tutorial so that you can go off-grid without any hassle. Let's get it started!

Chapter 1 – Getting Started

Before we let you know how to create your own hydro power system, it is important to cover all the basics. After all, if you truly want to run your power system in the future, then you got to know the principle behind it as well.

Hydro power (which is also known as water power at times) is the power generated from the movement of running water. For years, humans have been using it by building different kinds of watermills. This renewable source of energy can be used in several ways and help you generate power even in an adverse condition.

Unlike other renewable sources of energy (like solar or wind power), hydro power doesn't have plenty of requirements. For instance, one can generate solar power only when the sun is out. Even if you don't live in an area which has an abundant source of water, you can create an artificial stream as well. Afterward, you can back-feed the same water again to the system and use it time after time to generate power.

Even in ancient Roman or Han Empire, running water was used for different purposes. Though, in the modern times, we have started building dams to generate abundant of hydro power. Plenty of times, this has a negative impact on the environment. There are different ways to harness the power of the water. Some of them are as follows:

- *Run-of-the-river power*: It harnesses the kinetic energy of rivers and streams by building dams.

- *Small and micro hydro projects*: They are ideally used in villages and small towns. They can generate from a few hundred to as much as 10 megawatts of power.

- *Pumped-storage power*: It uses a pump to transfer water from one source to another and generates power from its movement.

- Pressure buffering technique: In this, water is pumped to generate power and is sent back to the reservoir.

Most of the micro hydro power systems that are created to meet the demands of a single house are of the following types.

- *Stand-alone with batteries*: Just like an off-grid wind or solar system, it utilizes a renewable source of energy to charge generators or a battery bank. Additionally, you need to install a controlling device that will protect the batteries from over-charging. These batteries can be used to provide power to the main grid of your house.

- *Battery tied with the grid*: This is just the modified version of the above system. In this, you will rely on the main grid whenever there is a shortage of hydro power. In the case of emergency, the grid will provide power to the battery. Since your hydro power system will already be transferring the extra amount of power, it will be subtracted from the total consumption. This is known as net metering. It is already used by solar and wind power plants and can reduce your electricity bill to a great extent without facing any blackout.

- *Grid-tied battery-less system*: In this, the power that is produced by the turbine directly goes to the utility line. Since there is no battery bank in between, an inverter is placed to convert DC to AC. Since a hydro power generator produces DC (Direct Current) electricity and our modern household appliances use AC (Alternating Current), an inverter is installed in between to make the conversion. Nevertheless, this system has some major drawbacks, as it can't store power in batteries or produce back-feed current.

You should try to implement a technique similar to that of pressure buffering as it doesn't cause any harmful effect to the environment. Also, try to join your system to the grid in order to obtain productive results.

How is hydropower generated?

This might surprise you, but more than 15% of the world power need is fulfilled by water. If you wish to create your own hydro power system, then you should understand the basic principle of it as well.

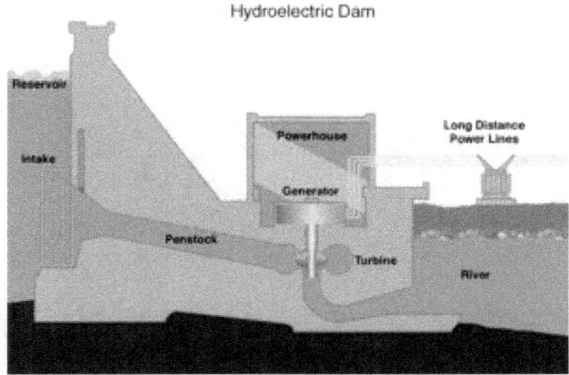

In a hydro power system, the kinetic energy of the water is used to generate electricity. The system includes a turbine, which is in contact with the stream of running water. As the running water moves the shaft of the turbine, it converts the kinetic energy of the water to mechanical energy, which is used to generate power.

In most of the cases, water is stored at a higher surface in order to maintain its high speed. It is ideally passed through a vessel or a gate, which is mostly known as penstock. The generator works on the age-old principle of magnets. That is, when you pass a magnet to a conductor, it starts flowing electric current.

The rotor and stator are the basic part of a generator. The field pole rotates at a respective speed in a rotor. As it rotates, it passes the pole through the stator to cause the identical electric field effect. Therefore, the flow of the water should be at a constant speed. This will ensure the optimized production of the electricity.

Great! Now when are familiar with the basic principle of hydro power, we can proceed to the next section. Read on and learn how you can create your own hydro power system and go off-grid.

Chapter 2 – Micro Hydro System Sizing & Components

If you wish to go off-grid and be self-sustained, then you need to know how much power your family needs. Before we provide an in-depth and stepwise tutorial to create your own hydro power system, it is important to be familiar with all the major components of it as well as its sizing.

Hydro power system sizing

Ideally, if you have a constant flow of water, then you don't need to make any added efforts. Otherwise, you can create an artificial system by placing a water unit at a significant height and letting it flow towards the turbine. The turbine would be connected to a 12, 24, or 48 volt batteries in the end.

Before starting the assembly, you need to know how much power your house needs. To start with, you need to scale the watt/hour capacity of every major electronic appliance. For example, an average microwave runs on 900 watts. Ideally, its hours/day ratio is 1/4. Therefore, its watts hour/day unit would be 225.

You can do the same for all the other major appliances like toasters, air conditioners, vacuum cleaners, fans, and more. Now, take a sum of all these units. This will be the total watts hours/day requirement of your house. Your hydro power system has to generate an enough amount of constant energy to meet your household needs.

Most of the times, the energy produced by a hydro power system is more than the average hourly need of a house. In that case, you can connect your generator to batteries and can use this energy afterward.

Components of a hydro power system

It is of utmost importance to be aware of all the major components of a hydro power system. This will make it easier for you to create the entire system and bring everything together in no time. Following are some of the major components of a micro hydro power system that you should focus on.

Intake

Needless to say, this is one of the most important parts of the system and you should try to install it beforehand. It is the part of the stream from where the penstock begins. To get best results, try to attain a direct flow from the pipeline to the stream. If the flow is not strong enough, then you can also choose to narrow the stream or add rocks or other organic materials to enhance its capacity. Though, the place where the stream drops should be stable in nature and not in touch with debris or any contaminated product.

Penstock and pipes

It is also known as the fuel line, as it provides the fuel (water) for the system to work. While working on this, you need to make sure that the pipes are of an ideal length and won't cause any loss of water. It should not be either too large or too small. Most of the times, they are made of either PVC or steel. The diameter of your pipe solely depends on the site and the amount of electricity you wish to produce. Budget and friction loss also determines its quality.

Turbine

There are different kinds of turbines that are readily available in the market. They work as a center engine for the entire system. You might have to consider your nozzle shape and the flow of the water in mind while buying a turbine. They also include a wheel (runner) that takes the kinetic energy of the water and use it to turn the shaft, which generates power in the end.

Controls

Just like any other system, it also needs regulators and controllers. With the help of a regulator, you can control the amount of the current that would be passed to the dump load. Since the turbine won't be disconnected from the batteries, there should be a way to control its movement as well. The over-speeding of turbines might damage it in the long run. Even if you are installing a grid-tied or a battery-less system, you would need controls to monitor the flow of current.

Dump load

It is also known as a diversion or shunt load. Ideally, it is the kind of electrical resistant heater that is installed to optimize the turbine's function. If the batteries can't accept the overload produced by the turbine, then it goes to a dump load (which is an air or water heater) in order to shunt it.

Storage battery

Your system needs a way to store the amount of energy that is produced by the turbine. These batteries are connected to the grid and supplies essential power to meet your household needs. If you have a large power system, then you can come up with a dedicated battery bank in order to store the amount of produced energy.

Inverter and DC connection

At the end of the battery bank, a connection needs to be made from its DC output to inverter's input. Ideally, there is a DC connector placed in between that allows the inverter to be taken out whenever needed (for servicing or replacement). It also protects a wiring fault between both the units. As stated, an inverter is used to convert DC into AC electricity. It is transferred to the main grid (if you have a grid-tied system) and is often connected to a backup generator as well.

Metering

Unlike other renewable sources of energy like solar or wind power generators, one needs to constantly monitor the output of hydro power generators. In order to get immediate readings, you need to install tools like amp-hour meter, watt-hour meter, and battery meter. It will let you know the status of your system, so that you can take care of its operation.

Additionally, you need to install a utility meter as well. It is also known as a KWH meter (Kilowatt-Hour meter) and keeps a track of how much electricity you are consuming from the grid. This will help you quantify the overall performance of your system in the long run.

Main panel

The AC breaker panel is ideally a wall-mounted box that is connected to the main circuit of the house. There are additional breakers that are installed, which can be disconnected anytime to prevent electric fires. It is connected to different rooms (and plugs) of the house at the end.

Performance of the system

The performance of your water system will depend on the above-listed components. Ideally, the greater the pressure would be at the nozzle, the more amount of power it would produce. You can always take the assistance of a pump or natural elevation to boost its pressure.

The volume of water also matters a lot. You should measure the amount of gallons per minute your system would receive. Try to reach an optimum solution in order to save water while generating maximum power from it. It has been discovered that sites with a good elevation, few nozzles, and smaller pipes not only costs less, but produces optimum results as well.

You are almost there. Now when you know about all the major components of a hydro power system and the basic working principle of it, you can easily build one on your own as well. As stated, there are different kinds of hydro power systems that can be created. We will provide tutorials for different types of power generators that you can easily build in no time to go off-grid.

Chapter 3 – Micro Hydro Power Generator

In order to start with, we will make you familiar with a basic hydro power generator. You can create this by using simple objects and utilize the kinetic energy of running water to generate power.

The principle behind this technique is almost similar to that of the windmills. In this, running water will rotate the shaft of the turbine in order to generate electricity. As already stated, we will be taking the assistance of magnets and coils to achieve the same.

In order to go off-grid and create your own mini hydro power system, you need to follow these steps.

Start by working on the disks

The first step is to create the disks for the generator. You need to know the basic idea of how the generator is going to be formed. It will be divided into two major parts – rotor and stator.

As the name suggests, stator would be the stationary part of the unit and would include various coils. The rotor would be the second part of the system and will include different magnets. These magnets will move over the coil region and will induce current after their movement.

In order to start with, work on the design of the disk. You can either use cardboard, metal, or even plastic to make the disk. The choice of metal depends on your requirements. Cut the material and attach it to the main unit. Now, make a hole right at the center of the stator. If you don't have enough space, then go for a 1 or 2 cm wide measurement else you can go for as wide as a 5 cm hole.

Stator

After making a base for the disk, the next step is to prepare the stator. Ideally, you need to attach four coils to the stator unit. To increase their surface area, try to make oval sections in the unit. Simply take a metal wire (could be a copper wire as well) and start turning it across the stator unit. You can make as many as a few hundred turns. You need to follow the same drill again until you have four different coils of the same kind.

You can also take the assistance of the above illustration and arrange the coil in the same fashion. While winding the coil to the material, you need to make sure that they would be alternate in nature. That is, if the first coil is being rotated in the clockwise direction, then second should be in anti-clockwise and vice-versa. This will make sure that the electrons will follow a designated path.

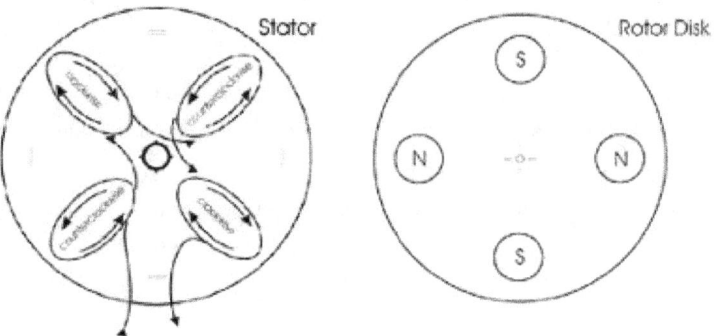

Now, give it a final touch and bring everything together using an adhesive or an insulating tape. Use a multimeter to make sure the amount of resistance is ideal in nature. You should get a reason of 10 Ohms if everything is ideally placed.

Rotor

The next part after finishing the stator would be working on the rotor. As you know this would be the moving part of the unit. Creating a rotor would be quite easy. Simply take the identical part of the stator and attach four magnets to it.

They would also be placed in an alternate manner – that is, first place the south pole of the magnet, then the north pole of the second magnet, and so on. This will maintain the alternative polarity (N-S-N-S).

Turbine

If you are working on a small unit, then you can simply use cork and forks to create the wings of the turbine. You can always buy a readily available turbine from the market as well. Nevertheless, if you wish to make your own moving unit, then simply take equal size measurements and cut the wings of the turbine from the base material.

Bring all these units together to assemble the turbine. Take the cork (at least 10 mm deep), which will act as a base of the wings of the turbine. Assemble these units to the base with the help of an adhesive. Make sure that they are placed in a precise manner and maintain an equal distance between each unit.

Making it work

Great! You are almost there. Now is the time to put everything together and make the entire system work. You need to fix your unit in a container. Additionally, you need to make sure of the stream of water. In order to do so, place a water tank at a higher point and build a stream of water from it. Also, try to link this stream to another container. Later, you can use this water for any other household purpose or can simply apply a feedback mechanism to it as well.

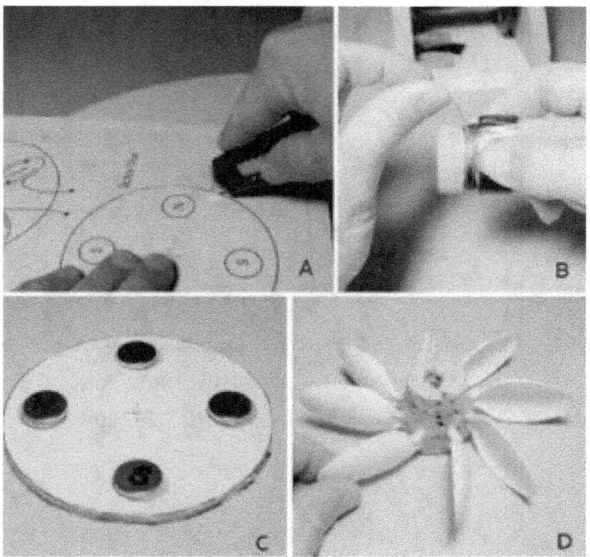

Additionally, make holes right in the center of the tank of uniform width. While doing so, you need to ensure that the coil is placed above the hole and won't get in touch with the running water. Place the magnets together as well and ensure that the separation distance is not more than 3-4 mm between them.

Lastly, place the wings carefully as well, letting them face the neck of the bottle. Your generator is now almost ready. After running the water, the wings of your turbine should get rotated by the stream. This will move the shaft of the generator and produce power to charge the connected batteries.

Now when you are able to build a mini hydro power generator, let's take a step up and work on a paddle wheel hydro generator.

Chapter 4 – Paddle Wheel Hydro Generator

By creating a paddle wheel hydro power generator, you can easily harness the power of running water and use it to generate electricity. Fortunately, it is quite easy to create this generator. You can get enough power from this system to run a few appliances or power a room or two. In future, you can either scale it up or simply create 2-3 generators of this kind to power every room separately.

This is a basic power system in which PVC pipes and a pump has been used. The pump is used to supply water to the turbine. As it rotates the blades of the propeller, it generates electricity. The water is collected at the base and is used again to move the turbine.

The basics

In order to apply maximum force to the turbine blades, you need to install your system in such a way that the water will enter it in a radial direction. It should not run parallel (or even perpendicular) to it. While working on this system, you need to make sure that your energy is not lost in the process.

You can always take the assistance of gravity in order to run the water. If you think that the pump is consuming a lot of energy in the system and that the amount of net energy produced is not enough, then you can always remove it. Instead, you can place the water unit at an elevated place and use columns to drive the flow of the water.

While designing this one, make sure that the water will enter in a radial way. This would be possible by joining PVC pipes together and attaching a radial nozzle at the end. To work on a mini paddle wheel based hydro power system, you can take 5 ft 2 inch pipes.

Water flow

In this system, the direction of water flow is of utmost importance. You need to pay an extra attention to the placement of turbine as well. Needless to say, before working on the entire structure, you need to pay close attention to the direction of water flow.

After installing a base unit where water would be stored, simply connect the pipes and install the radial nozzle. This would supply water tangent to the face of the turbine. You have to measure the spin of the propeller in a timely manner. In order to optimize the result, you might have to change the direction and stick with the one that would spin the blades of the turbine in the best way.

Additionally, you need to maintain a little distance between the end of the nozzle and the turbine. If it is too close, then it might not have enough pressure to move the blades. In order to maximize its impact, place it at a distance in order to boost its pressure in a natural way.

Building the remaining structure

You can alter the structure as per your needs. One of the best parts of the paddle structure is that it is portable in nature. Also, you can create a separate structure for every room individually as well. Start by creating a basic wooden structure. Though, you can use any other material as well (like metal or plastic), but a wooden structure would be easier to build.

On the top of the structure, place a water container. The size of the container will depend on the surface area of the structure. Make a small opening at the bottom of the container and join it with PVC pipes. Connect as many pipes as you want and run them through the entire structure.

Make sure that you have an enclosed structure, so that there won't be any splashing or wastage of water. Place the turbine at the bottom of the container and make sure that it is aligned with the rest of the structure. The other end of the turbine would be connected to the generator.

Additionally, you need to make sure that the entire structure has a well-defined and isolated base to store the remaining amount of water. You can install a pump here to back-feed this water to the container again or can simply store it as well.

Making it work

In the end, you need to show your plumbing skills to make the entire system work. Ideally, you can go with a 2-inch feed line and introduce necessary bends in between. You can get these T-shaped bends readily from a hardware store. There are also some 90-degree bends that you need to install in the system. We highly recommend using PVC lines in the system, as they are quite easy to install and modify.

There is no ideal number of bends that you need to implement in the system. After putting everything together, simply test it using a multimeter and keep improvising to get best results. Place the system wherever you want and supply its output to the grid.

Chapter 5 – Recycled Hydro Power Generator

This might surprise you, but you can create a hydro power generator from simple household objects as well. This is why, the system is known as a recycled power generator. In this system, we will be using a bicycle wheel as a water turbine and other basic household tools. If you are running low on budget, then this is just an ideal option for you.

By using minimum materials, this water turbine can be created and generate electricity in the end. Most of the bicycle and automotive parts have been used in this system. Additionally, it won't take a lot of time to build this one. To start with, you would need to following tools and materials.

- An old bicycle
- Drill and screws
- Generator
- Screw driver
- Plastic cups
- Welder
- Saw

Procedure

After collecting all the above-mentioned materials and tools, you can simply follow these basic instructions in order to create your own recycled hydro power generator.

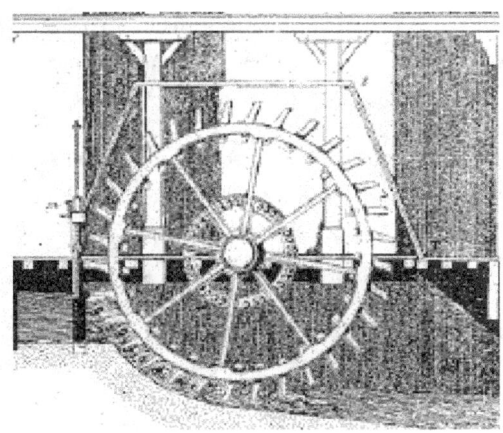

In order to start with, remove the front wheel of an old bicycle. In order to do so, you need to turn the axle nut and move it in an anti-clockwise direction. Additionally, you need to remove the paddle chain as well. If you haven't done this before, then simply move the paddles in the opposite direction and push the chain towards your side in order to free it from the bicycle.

If you have an old car, then you can remove its alternator in order to use it as this system's generator. Nevertheless, you can always buy a new one as well. Now, welt the generator in such a way that the chain would be wrapped around its pulley, when it is centered.

Get rid of the pulley with gear sprockets. If you can't do this, then simply weld the sprocket to the rightly centered position. After when the chain would be secured, it should be able to turn the rear wheel of the generator.

Great! You are almost there. Raise the bike seat to its maximum level. This can be done by taking the assistant of a screw that is located near the bike seat. At the same time, sew at least a dozen of plastic cups together.

Start by sewing the series of plastic cups to the rear wheel of the bicycle. Try to space them evenly with a gap of around 2-inches in between. Make sure that they all would be placed in the same direction. Ideally, they should be placed in clockwise direction to go well with the alternator.

At the same time, prepare a flowing creek of water. If you already have a natural source or a running stream of water, then you don't have to make any added efforts. Otherwise, you need to create an artificial stream to make it work. Place the system into the stream of running water safely. While doing so, make sure that the generator would not get submerged in water.

As you would place the system, you will see that the bike is upside down and the generator is above the water level. Also, the plastic cups should be facing the running water. They would be able to push the wheel of the bicycle after being in touch with running water.

If the flow of the water is strong enough, then the wheel will start turning and generate power. Simply connect the generator to a battery bank in order to store this energy. Additionally, if the plastic cups are getting damaged or are not able to withstand the flow of the water, then simply cut plastic balls in half and use them instead as they are more durable in nature.

That's it! The system will start working in no time. You can make desired changes in it as well in order to optimize it.

Now when you are familiar with different kinds of hydro power systems, nothing can really stop you anymore. Go ahead and start building your preferred system and go off-grid without any trouble.

Conclusion

Congratulations on completing the book so soon! We are sure you must have had a great time learning how to build a hydro power system of your own. If you wish to go off-grid and live a self-sustaining lifestyle, then you should start from the basics. Even after installing a solar or wind power plant, you should make an effort in order to harness energy from running water.

Unlike most of the other renewable sources of energy, hydro power doesn't come with any restrictions. You can either use a natural source of running water or can create one on your own as well. It is highly flexible in nature, which makes it so efficient.

In this guide, we first made you familiar with all the basics of hydro power generation. We know that without knowing the core principle, it can be tough to create an entire power system. From the basic parts of a hydro power system to the working principle of it, we have covered it all.

Furthermore, to make things easier for you, we came up with three different ways to create hydro power generators. Depending on your requirements and budget, you can build the power system of your choice. Ideally, you should start with the easiest one and then can evolve it with time.

A stepwise tutorial has been provided for all the three power systems. This will make it easier for you to go off-grid and build your own hydro power generator. Don't wait anymore and harness the power of the water while creating these highly efficient power systems.

FREE Bonus Reminder

If you have not grabbed it yet, please go ahead and download your special bonus report *"DIY Projects. 13 Useful & Easy To Make DIY Projects To Save Money & Improve Your Home!"*

Simply Click the Button Below

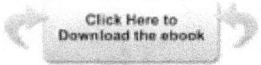

OR **Go to This Page**

http://diyhomecraft.com/free

BONUS #2: More Free & Discounted Books or Products

Do you want to receive more Free/Discounted Books or Products?

We have a mailing list where we send out our new Books or Products when they go free or with a discount on Amazon. Click on the link below to sign up for Free & Discount Book & Product Promotions.

=> Sign Up for Free & Discount Book & Product Promotions <=

OR Go to this URL

http://bit.ly/1WBb1Ek

www.ingramcontent.com/pod-product-compliance
Lightning Source LLC
Chambersburg PA
CBHW030108230526
45471CB00003B/1309